🌟 Welcome to Your 4th Grade Math Adventure! 🌟

Dear Student,

You are about to begin an exciting journey through 4th grade math! This workbook was created just for YOU — to help you build confidence AND develop strong math skills.

🧠 Why Practice Matters

Math is like a muscle — the more you use it, the stronger it gets! Practicing helps you:

- Master important skills like multiplication, division, and place value
- Solve real-world problems with confidence
- Prepare for quizzes, tests, and success in the next grade
- Build **focus, discipline, and problem-solving** superpowers

💡 What You'll Learn

This workbook covers:

- Numbers and place value up to one million
- Rounding, comparing, and writing numbers
- Adding, subtracting, multiplying, and dividing large numbers
- Understanding number patterns and solving multi-step word problems

🙌 You Can Do This!

Some problems will be easy. Others will challenge you. That's okay! Every page you complete is a victory. Don't be afraid to make mistakes — they are part of learning. Take your time, use scratch paper, and don't forget to **ask questions** when you need help.

🎯 Our Goal

To help you become a confident problem-solver, critical thinker, and math rockstar!

Comparing Numbers

- Always compare from **left to right** (largest place value first).
- Use comparison symbols: > (greater) < (less) = (equal)
- Steps: Line up digits, compare each from left to right.

✓ Example: ... 01 > ... 449

...ing & Dividing Using Place Value

✕ MULTIPLICATION:
...RTIAL PRODUCTS, OR DISTRIBUTIVE PROPERTY.
...3 × 4 = (20×4) + (3×4) = 80 + 12 = 92
...(200×6) + (10×6) + (3×6) = 1,278
÷ DIVISION:
...OR LONG DIVISION STRATEGIES.
... CHECK FOR REMAINDERS.
...E: 483 ÷ 3 = 161

Rounding Numbers

- Find the place you're rounding to ...
- Look at the digit to the right ...
- 5 or more: Round up ...
- 4 or less: Keep the same and change digits ...
... rhyme "5 or more, raise the score; 4 or less ..."

✓ Examples:
- 3,478 to nearest hundred → 3,5...
- 62,849 to nearest thousand → 63,...

Adding & Subtracting Multi-digit Numbers

- Line up by place value and add/subtract from right to left.
- Regroup (borrow/carry) as needed.

✓ Addition Example: 3,486 + 2,754 = 6,240

✓ Subtraction Example: 7,003 − 4,768 = 2,235

Reading & Writing Numbers

- Place value tells the value of a digit based on its position.
- Use commas to separate periods: Ones, Thousands, Millions (e.g., ...356,742).
- **Standard Form**: 124,506
- **Word Form**: One hundred twenty-four thousand, five hundred six
- **Expanded Form**: 100,000 + 20,000 + 4,000 + 500 + 6
- Tip: Say the number out loud to help with commas ...

PRO TIPS

- Label your answers (e.g., books, miles).
- Estimate first to check if your answer makes sense.
- Show your work clearly to avoid mistakes.

4th Grade Math Workbook

Workbook Index

☑ Student Goals & Progress Checklist

Instructions: Check off each box after you complete the section. You can also write in the date you completed it!

Section	Completed	Date	Parent/Teacher Initial
Examples & Step-by-Step Practice	☐	________	________
Pre-Test	☐	________	________
Read & Write Numbers (1–20)	☐	________	________
Compare Numbers (21–40)	☐	________	________
Rounding Numbers (41–60)	☐	________	________
Addition & Subtraction (61–80)	☐	________	________
Multiplication Using Place Value (81–100)	☐	________	________
Multi-Step Story Problems (1–25)	☐	________	________
Post-Test	☐	________	________
Certificate Earned! 🎓	☐	________	________

4th Grade Math Workbook

Number & Operations in Base Ten Examples

Reading and Writing Numbers

Example: Write 654,321 in standard form.

Step 1: Read the number aloud: six hundred fifty-four thousand, three hundred twenty-one.

Step 2: Write the digits without words or commas: 654321.

Answer: 654,321

Comparing Numbers

Example: Compare 789,123 and 789,321.

Step 1: Start comparing from the leftmost digits: both start with 789.

Step 2: Compare the hundreds: 1 (in 789,123) vs 3 (in 789,321) → $1 < 3$.

Answer: 789,123 < 789,321

Rounding Numbers

Example: Round 834,291 to the nearest thousand.

Step 1: Look at the thousands place: 4 (in 834,291).

Step 2: Look at the hundreds place to decide: 2 → round down.

Answer: 834,000

Adding and Subtracting Multi-digit Numbers

Example: Add 24,567 + 13,982.

Step 1: Line up the numbers by place value.

Step 2: Add from right to left:

7+2=9, 6+8=14 (write 4, carry 1), 5+9+1=15 (w[illegible] 4+3+1=8, 2+1=3

Answer: 38,549

Multiplying and Dividing Using Place Value

Example: Multiply 312 × 4 using place value.

Step 1: Break 312 into parts: 300 + 10 + 2

Step 2: Multiply each part: 300×4=1200, 10×4= [illegible]

Step 3: Add the products: 1200 + 40 + 8 = 1248

Answer: 1248

4th Grade Math – Pretest

Pretest

1. Write 562,713 in standard form.

2. Compare using <, >, or =: 456,298 ____ 456,982

3. Round 382,649 to the nearest thousand.

4. Add: 31,475 + 24,316

5. Subtract: 85,000 – 27,684

6. Multiply: 312 × 4 using place value

7. Round 918,275 to the nearest ten thousand.

8. Write the number 720,130 in standard form.

9. Compare using <, >, or =: 803,452 ___ 803,425

10. Multiply: 153 × 6 using place value

4th Grade Math – Problems 1-100 – You Got This!

1. Read and write this number in standard form and write in word form: 377,240

2. Read and write this number in standard form and write in word form: 243,185

3. Read and write this number in standard form and write in word form: 32,442

4. Read and write this number in standard form and write in word form: 413,943

5. Read and write this number in standard form and write in word form: 553,880

6. Read and write this number in standard form and write in word form: 896,761

7. Read and write this number in standard form and [illegible] form: 513,835

8. Read and write this number in standard form and [illegible] form: 195,875

9. Read and write this number in standard form and [illegible] form: 550,744

10. Read and write this number in standard form and [illegible] form: 153,352

11. Read and write this number in standard form and [illegible] form: 242,366

12. Read and write this number in standard form and [illegible] form: 234,554

13. Read and write this number in standard form and [illegible] form: 717,267

14. Read and write this number in standard form and write in word form: 788,219

15. Read and write this number in standard form and write in word form: 293,043

16. Read and write this number in standard form and write in word form: 921,356

17. Read and write this number in standard form and write in word form: 611,636

18. Read and write this number in standard form and write in word form: 894,002

19. Read and write this number in standard form and write in word form: 439,150

20. Read and write this number in standard form and write in word form: 737,686

21. Using greater than or less than symbols, < or >, Compare: 33541 and 301915

22. Using greater than or less than symbols, < or >, Compare: 443377 and 655375

23. Using greater than or less than symbols, < or >, Compare: 778838 and 535170

24. Using greater than or less than symbols, < or >, Compare: 890779 and 688052

25. Using greater than or less than symbols, < or >, Compare: 141026 and 22415

26. Using greater than or less than symbols, < or >, Compare: 893732 and 956678

27. Using greater than or less than symbols, < or >, Compare: 123421 and 990104

28. Using greater than or less than symbols, < or >, Compare: 497433 and 505530

29. Using greater than or less than symbols, < or >, Compare: 262370 and 583123

30. Using greater than or less than symbols, < or >, Compare: 127392 and 15001

31. Using greater than or less than symbols, < or >, Compare: 255270 and 917925

32. Using greater than or less than symbols, < or >, Compare: 318256 and 982393

33. Using greater than or less than symbols, < or >, Compare: 453493 and 78600

34. Using greater than or less than symbols, < or >, Compare: 21050 and 638213

35. Using greater than or less than symbols, < or >, Compare: 492898 and 825321

36. Using greater than or less than symbols, < or >, Compare: 877409 and 977812

37. Using greater than or less than symbols, < or >, Compare: 798147 and 656418

38. Using greater than or less than symbols, < or >, Compare: 106188 and 183813

39. Using greater than or less than symbols, < or >, Compare: 169010 and 879742

40. Using greater than or less than symbols, < or >, Compare: 854429 and 530529

41. Round 617159 to the nearest ten thousand.

42. Round 739617 to the nearest ten.

43. Round 625255 to the nearest ten.

44. Round 457795 to the nearest ten thousand.

45. Round 25314 to the nearest hundred.

46. Round 580779 to the nearest ten thousand.

47. Round 207526 to the nearest ten.

48. Round 233467 to the nearest ten.

49. Round 67924 to the nearest thousand.

50. Round 528016 to the nearest thousand.

51. Round 601359 to the nearest hundred thousand.

52. Round 689174 to the nearest hundred.

53. Round 112543 to the nearest hundred thousand.

54. Round 538792 to the nearest thousand.

55. Round 188133 to the nearest hundred thousand.

56. Round 941901 to the nearest hundred thousand.

57. Round 42070 to the nearest thousand.

58. Round 456571 to the nearest hundred.

59. Round 441518 to the nearest ten thousand.

60. Round 745523 to the nearest ten thousand.

61. Subtract: 56424 - 2966

62. Add: 48323 + 5512

63. Subtract: 51707 - 3076

64. Add: 80710 + 4502

65. Subtract: 74706 - 1482

66. Add: 70194 + 8198

67. Subtract: 70265 - 3083

68. Add: 65944 + 7183

69. Subtract: 55440 - 1142

70. Add: 77263 + 8369

71. Subtract: 80607 - 2183

72. Add: 86928 + 3490

73. Subtract: 86637 - 1068

74. Add: 63066 + 2387

75. Subtract: 96161 - 3342

76. Add: 66709 + 2833

77. Subtract: 78231 - 1828

78. Add: 92249 + 8731

79. Subtract: 66617 - 1659

80. Add: 12268 + 7423

81. Multiply using place value: 772 × 3

82. Multiply using place value: 362 × 3

83. Multiply using place value: 651 × 6

84. Multiply using place value: 935 × 2

85. Multiply using place value: 520 × 6

86. Multiply using place value: 689 × 2

87. Multiply using place value: 376 × 7

88. Multiply using place value: 805 × 7

89. Multiply using place value: 568 × 5

90. Multiply using place value: 588 × 5

91. Multiply using place value: 817 × 4

92. Multiply using place value: 197 × 2

93. Multiply using place value: 113 × 5

94. Multiply using place value: 869 × 3

95. Multiply using place value: 785 × 7

96. Multiply using place value: 588 × 4

97. Multiply using place value: 184 × 8

98. Multiply using place value: 201 × 8

99. Multiply using place value: 650 × 7

100. Multiply using place value: 466 × 8

25 Multi-Step 4th Grade Math Story Problems

1. Kalel collected 407863 stamps and Whitley collected 481720 stamps. Who collected more? By how much?

2. The population of the town Boggle-bee-bear is 38723. What is the population rounded to the nearest thousand and nearest ten thousand?

3. A factory produced 29176 toys in January and 27447 toys in February. If 6082 toys were defective and thrown out, how many good toys were produced? If each toy cost 1.25 dollars to make, how much money did the factory lose?

4. Each school supply box contains 571 pencils. If a school orders 6 boxes, how many pencils will they receive in total? 10% of the pencils they received in total were defective. How many good pencils are there?

5. Evelynn has 578 new books to arrange evenly on 5 shelves. How many books will be on each shelf? Will any be left over?

6. Sophia collected 225728 stamps and Gabriella collected 226353 stamps. Who collected more? By how much?

7. The population of the town Snicker-dee-doodle is 58405. What is the population rounded to the nearest thousand and nearest ten thousand?

8. A factory produced 48744 pop-it toys in January and 12274 toys in February. If 9056 pop-it toys were defective and thrown out, how many good pop-it toys were produced?

9. Each school supply box contains 603 markers. If a school orders 2 boxes, how many markers will they receive in total?

10. A bank teller has 583 stacks of money to arrange evenly on 7 shelves. How many stacks of money will be on each shelf? Will any be left over?

11. Harbor collected 459944 energy drinks and Aubrey collected 314682 energy drinks. Who collected more? By how much?

12. The population of dogs in the town of Ruff-Woof-Ruff is 99618. What is the population of dogs rounded to the nearest thousand and nearest ten thousand?

13. A factory produced 452 jelly-cats in April that sold [illegible] dollars each and 351 jelly-cats in May that were limited edition and sold for 200.00 each. [illegible] money did the factor make in the month of April and May? Which month did they make the [illegible]

14. Each lunch box contains 3 mini crumble cookies. If a [illegible] 5 boxes, how many mini crumble cookies will they receive in total?

15. A grocery worker has 861 boxes of cereal to arrange [illegible] shelves. How many boxes of cereal will be on each shelf? Will any be left over?

16. Bob collected [illegible]7502 toy cars and Victor collected [illegible] cars. Who collected more? By how much?

17. The population of cats in a town is 24030. What is the population rounded to the nearest thousand and nearest ten thousand?

18. Cane's Fast-Food Chicken produced 42262 chicken tenders in January and 31082 chicken tenders in February. If 11044 chicken tenders were defective and thrown out, how many good chicken tenders were produced?

19. Each bag of strawberry seeds contains 451 seeds. If a school orders 6 boxes, how many seeds will they receive in total? If the garden they plant them has 12 pots, each that can take 125 seeds, how many strawberry seeds will be left over?

20. Your mom has asked you to organize all the cups in the pantry. If you have 602 cups to organize on 9 shelves, how many cups will go on a shelf? Will there be any left over?

21. Chelse found 876 toys that her children left out on the floor. She asked her children to pick up all the toys, but there were still 789 toys left on the floor when she went to check. How many toys did her children pick up?

22. The number of Tic Toc videos that Gabriella watched is 41282. What is the number of videos that Gabriella watched rounded to the nearest thousand and nearest ten thousand?

23. A paint factory produced 24089 cans of paint in August and 13693 cans of paint in December. If 9636 cans of paint were defective and thrown out, how many good cans of paint were produced? If each can of paint cost 2.50 cents to make, how much money did the Paint Factory lose?

24. Each year for school shopping, Vic gives each child 250 dollars to spend. If he has five children and each spend 75% of their allowed money. How much money does each child have left over?

25. Robbie is organizing his room. He has three baskets for his medals and 76 medals. How many medals will go in each basket? Will there be any left over?

4th Grade Ma[illegible]st-Test

Topic: Numbers & Operati[illegible]

Difficulty: Chall[illegible]

1. Write the number nine hundred two thousand, seventy-f[illegible] form.

2. Round 97426 to the nearest ten thousand.

3. Compare using >, <, or =: 100491 ____ 99410

4. What is the value of the digit 7 in the number 8743[illegible]

5. Add: 72938 + 18426

6. Subtract: 91023 - 4729

7. Multiply: 73 × 48

8. Estimate the difference between 85390 and 29456 by rounding both to the nearest thousand.

9. Solve this two-step word problem:

A toy company made 625 toys each day for 7 days. They shipped 3590 toys to a store. How many toys do they have left?

10. Divide: 935 ÷ 5. Show your remainder. – BONUS QUESTION***

Answer Key – Step-by-Step Solutions

Pretest Solutions

1. Write 562,713 in standard form

Step 1: Read the number: 562,713.

Step 2: Remove commas: 562713.

Answer: 562,713

2. Compare using <, >, or =: 456,298 ___ 456,982

Step 1: Compare from left to right: 456,298 < 456,982.

Answer: 456,298 < 456,982

3. Round 382,649 to the nearest thousand

Step 1: Look at the thousands digit: 2.

Step 2: Look at hundreds: 6 → round up.

Answer: 383,000

4. Add: 31,475 + 24,316

Step 1: Line up numbers:

31,475 + 24,316 = 55,791

Answer: 55,791

5. Subtract: 85,000 – 27,684

Step 1: Line up numbers:

85,000 - 27,684 = 57,316

Answer: 57,316

6. Multiply: 312 × 4 using place value

Step 1: Break 312 into 300 + 10 + 2.

Step 2: Multiply: 300×4=1200, 10×4=40, 2×4=8.

Step 3: Total = 1248

Answer: 1248

7. Round 918,275 to the nearest ten thousand.

Step 1: Ten thousand digit is 1.

Step 2: Thousand digit is 8 → round up.

Answer: 920,000

8. Write the number 720,130 in standard form.

Step 1: Already in standard form: 720,130.

Answer: 720,130

9. Compare using <, >, or =: 803,452 ____ 803

Step 1: Compare digits: 452 > 425.

Answer: 803,452 > 803,425

10. Multiply: 153 × 6 using place value

Step 1: Break 153 into 100 + 50 + 3.

Step 2: Multiply: 100×6=600, 50×6=300, 3×

Step 3: Total = 918

Answer: 918

Read and Write Numbers in Standard and Word Form (Problems 1-20)

1. Number: 377240

Step 1: Standard Form: 377240

Step 2: Word Form: Three hundred seventy seven thousand two hundred forty

2. Number: 243185

Step 1: Standard Form: 243185

Step 2: Word Form: Two hundred forty three thousand one hundred eighty five

3. Number: 32442

Step 1: Standard Form: 32442

Step 2: Word Form: Thirty two thousand four hundred forty two

4. Number: 413943

Step 1: Standard Form: 413943

Step 2: Word Form: Four hundred thirteen thousand nine hundred forty three

5. Number: 553880

Step 1: Standard Form: 553880

Step 2: Word Form: Five hundred fifty three thousand eight hundred eighty

6. Number: 896761

Step 1: Standard Form: 896761

Step 2: Word Form: Eight hundred ninety six thousand seven hundred sixty one

7. Number: 513835

Step 1: Standard Form: 513835

Step 2: Word Form: Five hundred thirteen thousand eight hundred thirty five

8. Number: 195875

Step 1: Standard Form: 195875

Step 2: Word Form: One hundred ninety five thousand eight hundred seventy five

9. Number: 550744

Step 1: Standard Form: 550744

Step 2: Word Form: Five hundred fifty thousand seven hundred forty four

10. Number: 153352

Step 1: Standard Form: 153352

Step 2: Word Form: One hundred fifty three thousand three hundred fifty two

11. Number: 242366

Step 1: Standard Form: 242366

Step 2: Word Form: Two hundred forty two thousand three hundred sixty six

12. Number: 234554

Step 1: Standard Form: 234554

Step 2: Word Form: Two hundred thirty four thousand five hundred fifty four

13. Number: 717267

Step 1: Standard Form: 717267

Step 2: Word Form: Seven hundred seventeen thousand two hundred sixty seven

14. Number: 788219

Step 1: Standard Form: 788219

Step 2: Word Form: Seven hundred eighty eight thousand two hundred nineteen

15. Number: 293043

Step 1: Standard Form: 293043

Step 2: Word Form: Two hundred ninety three thous[illegible]

16. Number: 921356

Step 1: Standard Form: 921356

Step 2: Word Form: Nine hundred twenty one thous[illegible]ed fifty six

17. Number: 611636

Step 1: Standard Form: 611636

Step 2: Word Form: Six hundred eleven thousand si[illegible] six

18. Number: 894002

Step 1: Standard Form: 894002

Step 2: Word Form: Eight hundred ninety four thous[illegible]

19. Number: 439150

Step 1: Standard Form: 439150

Step 2: Word Form: Four hundred thirty nine thousan[illegible]fifty

20. Number: 737686

Step 1: Standard Form: 737686

Step 2: Word Form: Seven hundred thirty seven thousand six hundred eighty six

Greater Than or Less Than Comparison (Problems 21-40)

21. Compare: 33541 and 301915

Step 1: Count the digits in each number.

- 33541 has 5 digits
- 301915 has 6 digits

Step 2: The number with more digits is greater.

Final Answer: 33541 < 301915

22. Compare: 443377 and 655375

Step 1: Count the digits in each number.

- 443377 has 6 digits
- 655375 has 6 digits

Step 2: Same number of digits, compare left to right.

- Compare digits: 4 vs 6

Final Answer: 443377 < 655375

23. Compare: 778838 and 535170

Step 1: Count the digits in each number.

- 778838 has 6 digits
- 535170 has 6 digits

Step 2: Same number of digits, compare left to right.

- Compare digits: 7 vs 5

Final Answer: 778838 > 535170

24. Compare: 890779 and 688052

Step 1: Count the digits in each number.

- 890779 has 6 digits
- 688052 has 6 digits

Step 2: Same number of digits, compare left to right.

- Compare digits: 8 vs 6

Final Answer: 890779 > 688052

25. Compare: 141026 and 22415

Step 1: Count the digits in each number.

- 141026 has 6 digits
- 22415 has 5 digits

Step 2: The number with more digits is greater.

Final Answer: 141026 > 22415

26. Compare: 893732 and 956678

Step 1: Count the digits in each number.

- 893732 has 6 digits
- 956678 has 6 digits

Step 2: Same number of digits, compare left to right.

- Compare digits: 8 vs 9

Final Answer: 893732 < 956678

27. Compare: 123421 and 990104

Step 1: Count the digits in each number.

- 123421 has 6 digits
- 990104 has 6 digits

Step 2: Same number of digits, compare left to right.

- Compare digits: 1 vs 9

Final Answer: 123421 < 990104

28. Compare: 497433 and 505530

Step 1: Count the digits in each number.

- 497433 has 6 digits
- 505530 has 6 digits

Step 2: Same number of digits, compare left to right.

- Compare digits: 4 vs 5

Final Answer: 497433 < 505530

29. Compare: 262370 and 583123

Step 1: Count the digits in each number.

- 262370 has 6 digits
- 583123 has 6 digits

Step 2: Same number of digits, compare left to right.

- Compare digits: 2 vs 5

Final Answer: 262370 < 583123

30. Compare: 127392 and 15001

Step 1: Count the digits in each number.

- 127392 has 6 digits
- 15001 has 5 digits

Step 2: The number with more digits is greater.

Final Answer: 127392 > 15001

31. Compare: 255270 and 917925

Step 1: Count the digits in each number.

- 255270 has 6 digits
- 917925 has 6 digits

Step 2: Same number of digits, compare left to right.

- Compare digits: 2 vs 9

Final Answer: 255270 < 917925

32. Compare: 318256 and 982393

Step 1: Count the digits in each number.

- 318256 has 6 digits
- 982393 has 6 digits

Step 2: Same number of digits, compare left to right.

- Compare digits: 3 vs 9

Final Answer: 318256 < 982393

33. Compare: 453493 and 78600

Step 1: Count the digits in each number.

- 453493 has 6 digits
- 78600 has 5 digits

Step 2: The number with more digits is greater.

Final Answer: 453493 > 78600

…21050 and 638213

…e digits in each number.

…5 digits

…s 6 digits

…mber with more digits is

…21050 < 638213

…492898 and 825321

…e digits in each number.

…has 6 digits

…s 6 digits

…umber of digits, compare

…gits: 4 vs 8

…492898 < 825321

…77409 and 977812

…e digits in each number.

…has 6 digits

…6 digits

…umber of digits, compare

…gits: 8 vs 9

…77409 < 977812

37. Compare: 798147 and 656418

Step 1: Count the digits in each number.

- 798147 has 6 digits
- 656418 has 6 digits

Step 2: Same number of digits, compare left to right.

- Compare digits: 7 vs 6

Final Answer: 798147 > 656418

38. Compare: 106188 and 183813

Step 1: Count the digits in each number.

- 106188 has 6 digits
- 183813 has 6 digits

Step 2: Same number of digits, compare left to right.

- Compare digits: 0 vs 8

Final Answer: 106188 < 183813

39. Compare: 169010 and 879742

Step 1: Count the digits in each number.

- 169010 has 6 digits
- 879742 has 6 digits

Step 2: Same number of digits, compare left to right.

- Compare digits: 1 vs 8

Final Answer: 169010 < 879742

40. Compare: 854429 and 530529

Step 1: Count the digits in each number.

- 854429 has 6 digits
- 530529 has 6 digits

Step 2: Same number of digits, compare left to right.

- Compare digits: 8 vs 5

Final Answer: 854429 > 530529

Rounding Numbers - (Problems 41-60)

41. Round 617169 to the nearest ten thousand

Step 1: Identify the digit in the ten thousand place.

- Digit in the ten thousand place: 1

Step 2: Look at the digit to the right of the ten thousand place to decide whether to round up or down.

- Digit to the right: 6

Step 3: Since 6 is 5 or greater, round up.

Final Answer: 620000

42. Round 739617 to the nearest ten

Step 1: Identify the digit in the ten place.

- Digit in the ten place: 1

Step 2: Look at the digit to the right of the ten place to decide whether to round up or down.

- Digit to the right: 6

Step 3: Since 6 is 5 or greater, round up.

Final Answer: 739620

43. Round 625255 to the nearest ten

Step 1: Identify the digit in the ten place.

- Digit in the ten place: 5

Step 2: Look at the digit to the right of the ten place to decide whether to round up or down.

- Digit to the right: 2

Step 3: Since 2 is less than 5, round down.

Final Answer: 625260

44. Round 457795 to the nearest ten thousand

Step 1: Identify the digit in the ten thousand place.

- Digit in the ten thousand place: 5

Step 2: Look at the digit to the right of the ten thousand place to decide whether to round up or down.

- Digit to the right: 4

Step 3: Since 4 is less than 5, round down.

Final Answer: 460000

45. Round 25514 to the nearest hundred

Step 1: Identify the digit in the hundred place.

- Digit in the hundred place: 5

Step 2: Look at the digit to the right of the hundred place to decide whether to round up or down.

- Digit to the right: 5

Step 3: Since 5 is 5 or greater, round up.

Final Answer: 25500

46. Round 530779 to the nearest ten thousand

Step 1: Identify the digit in the ten thousand place.

- Digit in the ten thousand place: 3

Step 2: Look at the digit to the right of the ten thousand place to decide whether to round up or down.

- Digit to the right: 5

Step 3: Since 5 is 5 or greater, round up.

Final Answer: 530000

47. Round 207526 to the nearest ten

Step 1: Identify the digit in the ten place.

- Digit in the ten place: 2

Step 2: Look at the digit to the right of the ten place to decide whether to round up or down.

- Digit to the right: 5

Step 3: Since 5 is 5 or greater, round up.

Final Answer: 207530

48. Round 233467 to the nearest ten

Step 1: Identify the digit in the ten place.

- Digit in the ten place: 6

Step 2: Look at the digit to the right of the ten place to decide whether to round up or down.

- Digit to the right: 4

Step 3: Since 4 is less than 5, round down.

Final Answer: 233470

49. Round 67924 to the nearest thousand

Step 1: Identify the digit in the thousand place.

- Digit in the thousand place: 7

Step 2: Look at the digit to the right of the thousand place to decide whether to round up or down.

- Digit to the right: 6

Step 3: Since 6 is 5 or greater, round up.

Final Answer: 68000

50. Round 528016 to the nearest thousand

Step 1: Identify the digit in the thousand place.

- Digit in the thousand place: 8

Step 2: Look at the digit to the right of the thousand place to decide whether to round up or down.

- Digit to the right: 2

Step 3: Since 2 is less than 5, round down.

Final Answer: 528000

...59 to the nearest
...nd
...the digit in the hundred

...undred thousand place: 6
...the digit to the right of the
...d place to decide whether
...own.
...ight: 0
...0 is less than 5, round

...00000

...9174 to the nearest

...the digit in the hundred

...undred place: 1
...the digit to the right of the
...o decide whether to round

...ight: 9
...9 is 5 or greater, round up.
...89200

53. Round 112543 to the nearest hundred thousand

Step 1: Identify the digit in the hundred thousand place.

- Digit in the hundred thousand place: 1

Step 2: Look at the digit to the right of the hundred thousand place to decide whether to round up or down.

- Digit to the right: 0

Step 3: Since 0 is less than 5, round down.

Final Answer: 100000

54. Round 538792 to the nearest thousand

Step 1: Identify the digit in the thousand place.

- Digit in the thousand place: 8

Step 2: Look at the digit to the right of the thousand place to decide whether to round up or down.

- Digit to the right: 3

Step 3: Since 3 is less than 5, round down.

Final Answer: 539000

55. Round 188133 to the nearest hundred thousand

Step 1: Identify the digit in the hundred thousand place.

- Digit in the hundred thousand place: 1

Step 2: Look at the digit to the right of the hundred thousand place to decide whether to round up or down.

- Digit to the right: 0

Step 3: Since 0 is less than 5, round down.

Final Answer: 200000

56. Round 941901 to the nearest hundred thousand

Step 1: Identify the digit in the hundred thousand place.

- Digit in the hundred thousand place: 9

Step 2: Look at the digit to the right of the hundred thousand place to decide whether to round up or down.

- Digit to the right: 0

Step 3: Since 0 is less than 5, round down.

Final Answer: 900000

57. Round 42070 to the nearest thousand

Step 1: Identify the digit in the thousand place.

- Digit in the thousand place: 2

Step 2: Look at the digit to the right of the thousand place to decide whether to round up or down.

- Digit to the right: 4

Step 3: Since 4 is less than 5, round down.

Final Answer: 42000

58. Round 456571 to the nearest hundred

Step 1: Identify the digit in the hundred place.

- Digit in the hundred place: 5

Step 2: Look at the digit to the right of the hundred place to decide whether to round up or down.

- Digit to the right: 6

Step 3: Since 6 is 5 or greater, round up.

Final Answer: 456600

59. Round 441518 to the nearest ten thousand

Step 1: Identify the digit in the ten thousand place.

- Digit in the ten thousand place: 4

Step 2: Look at the digit to the right of the ten thousand place to decide whether to round up or down.

- Digit to the right: 4

Step 3: Since 4 is less than 5, round down.

Final Answer: 440000

Addition and Subtraction - (Problems 61-80)

61. Subtract: 56424 - 2966

First, write the numbers one under the other, making sure the digits are lined up by place value:

```
  86928
+  3490
 --------
```

Now start adding from the right (ones place) and move to the left:

- 8 + 0 = 8, so the ones digit is 8.
- 2 + 9 = 11 — we write down 1 and carry over 1.
- 9 + 4 = 13, plus the 1 we carried makes 14 — write down 4 and carry 1.
- 6 + 3 = 9, plus 1 carried over makes 10 — write down 0 and carry 1.
- 8 + 0 = 8, plus 1 carried over makes 9.

So when we add it all up, we get:

☑ **Final Answer: 90,418**

62. Add 48,323 + 5,512

Step 1: Write the numbers one under the other, making sure the digits are lined up by place value:

```
 48323
 + 5512
 --------
```

Step 2: Start adding from the right (ones place) and move left:

1. 3 + 2 = 5 → Ones digit is 5
2. 2 + 1 = 3 → Tens digit is 3
3. 3 + 5 = 8 → Hundreds digit is 8
4. 8 + 5 = 13 → Write down 3 and carry over 1
5. 4 + 0 + 1 (carried) = 5 → Ten-thousands digit is 5

So when we add it all up, the final answer is:

☑ **Final Answer: 53,835**

63. Subtract 51707 - 3076

Step 1: Write the numbers one under the other, making sure the digits are lined up by place value:

```
  51707
- 3076
--------
```

Step 2: Start subtracting from the right (ones place) and move left:

- 7 - 6 = 1 → Ones digit is 1
- 0 - 7: Since 0 is less than 7, we borrow 1 from the next digit.
  10 - 7 = 3 → Tens digit is 3
- 6 - 0 = 6 → Hundreds digit is 6
- 1 - 3: Since 1 is less than 3, we borrow 1 from the 5 in the next column.
  11 - 3 = 8 → Thousands digit is 8
- 4 (borrowed from 5) → Ten-thousands digit is 4

So when we subtract, the final answer is:

Final Answer: 48,631

6… + 4502

…numbers one under the other, …digits are lined up by place

…dding from the right (ones …eft:

- … Ones digit is 2
- … ns digit is 1
- … Write down 2 and carry over 1
- … lus the 1 carried = 5 →
- … s 5
- … Ten-thousands digit is 8

… it all up, the final answer is:

…: 85,212

65. Subtract: 74706 - 1482

Let's solve it step by step:

First, write the numbers one under the other, lining up the digits by place value:

```
  74706
 - 1482
 --------
```

Start subtracting from the right (ones place):

6 - 2 = 4 → Ones place is 4.
0 - 8 → We can't subtract 8 from 0, so we borrow 1 from the tens column.
10 - 8 = 2 → Tens place is 2.
Now the tens column has 6 (borrowed from hundreds).
6 - 4 = 2 → Hundreds place is 2.
4 - 1 = 3 → Thousands place is 3.
7 - 0 = 7 → Ten-thousands place is 7.

So when we subtract, we get:

✅ Final Answer: 73,224

66. Add 70194 + 8198

Step 1: Write the numbers one under the other, making sure the digits are lined up by place value:

```
  70194
+  8198
--------
```

Step 2: Start adding from the right (ones place) and move left:

- 4 + 8 = 12 → Write down 2 and carry over 1
- 9 + 9 = 18, plus the 1 carried = 19 → Write down 9 and carry over 1
- 1 + 1 = 2, plus the 1 carried = 3 → Hundreds digit is 3
- 0 + 8 = 8 → Thousands digit is 8
- 7 remains → Ten-thousands digit is 7

So when we add it all up, the final answer is:

✅ Final Answer: 78,392

67. **Subtract 70265 - 3083**

Step 1: Write the numbers one under the other, making sure the digits are lined up by place value:

```
 70265
- 3083
 ---------
```

Step 2: Start subtracting from the right (ones place) and move left:

- 5 - 3 = 2 → Ones digit is 2
- 6 - 8: Since 6 is less than 8, we borrow 1 from the next digit.
  16 - 8 = 8 → Tens digit is 8
- 1 (borrowed from 2) - 0 = 1 → Hundreds digit is 1
- 0 - 0 = 0 → Thousands digit is 0
- 7 remains → Ten-thousands digit is 7

So when we subtract, the final answer is:

✅ **Final Answer: 67,182**

68. **Add 65944 + 7183**

Step 1: Write the numbers one under the other, making sure the digits are lined up by place value:

```
 65944
+ 7183
---------
```

Step 2: Start adding from the right (ones place) and move left:

- 4 + 3 = 7 → Ones digit is 7
- 4 + 8 = 12 → Write down 2 and carry over 1
- 9 + 1 = 10, plus the 1 carried = 11 → Write down 1 and carry over 1
- 5 + 7 = 12, plus the 1 carried = 13 → Write down 3 and carry over 1
- 6 + 1 (carried) = 7 → Ten-thousands digit is 7

So when we add it all up, the final answer is:

✅ **Final Answer: 73,127**

69. Subtract 55440 - 1142

Step 1: Write the numbers one under the other, making sure the digits are lined up by place value:

```
 55440
 - 1142
 ---------
```

Step 2: Start subtracting from the right (ones place) and move left:

- 0 - 2: We can't subtract 2 from 0, so we borrow 1 from the 4 (tens place).
  10 - 2 = 8 → Ones digit is 8
- 3 (after borrowing) - 4: Still can't subtract, so we borrow from the 4 in the hundreds place.
  13 - 4 = 9 → Tens digit is 9
- 3 - 1 = 2 → Hundreds digit is 2
- 5 - 1 = 4 → Thousands digit is 4
- 5 remains → Ten-thousands digit is 5

So when we subtract, the final answer is:

☑ **Final Answer: 54,298**

70. Add 77263 + 8369

Step 1: Write the numbers one under the other, making sure the digits are lined up by place value:

```
  77263
+ 8369
 ---------
```

Step 2: Start adding from the right (ones place) and move left:

- 3 + 9 = 12 → Write down 2 and carry over 1
- 6 + 6 = 12, plus the 1 carried = 13 → Write down 3 and carry over 1
- 2 + 3 = 5, plus the 1 carried = 6 → Hundreds digit is 6
- 7 + 8 = 15 → Write down 5 and carry over 1
- 7 + 1 (carried) = 8 → Ten-thousands digit is 8

So when we add it all up, the final answer is:

☑ **Final Answer: 85,632**

71. Subtract 80607 - 2183

Step 1: Write the numbers one under the other, making sure the digits are lined up by place value:

```
 80607
- 2183
-------
```

Step 2: Start subtracting from the right (ones place) and move left:

- 7 - 3 = 4 → Ones digit is 4
- 0 - 8: Since 0 is less than 8, we borrow 1 from the 6 (hundreds place).
  10 - 8 = 2 → Tens digit is 2
- 5 (after borrowing) - 1 = 4 → Hundreds digit is 4
- 0 - 2: Since 0 is less than 2, we borrow 1 from the 8 (ten-thousands place).
  10 - 2 = 8 → Thousands digit is 8
- 7 (after borrowing) → Ten-thousands digit is 7

So when we subtract, the final answer is:

Final Answer: 78,424

7[illegible] 8 + 3490

[illegible] numbers one under the other, [illegible] digits are lined up by place [illegible]

[illegible] adding from the right (ones [illegible] left:

- [illegible] Ones digit is 8
- [illegible] Write down 1 and carry over 1
- [illegible] plus the 1 carried = 14 → Write [illegible] carry over 1
- [illegible] plus the 1 carried = 10 → Write [illegible] carry over 1
- [illegible] plus the 1 carried = 9 → Ten-[illegible] is 9

[illegible] it all up, the final answer is:

[illegible]: 90,418

73. Subtract 86637 - 1068

Step 1: Write the numbers one under the other, making sure the digits are lined up by place value:

```
 86637
- 1068
---------
```

Step 2: Start subtracting from the right (ones place) and move left:

- 7 - 8: We can't subtract 8 from 7, so we borrow 1 from the 3 → 17 - 8 = 9
- 2 (after borrowing from the 3) - 6: We can't subtract 6 from 2, so borrow from the 6 → 12 - 6 = 6
- 5 (after borrowing) - 0 = 5
- 6 - 1 = 5
- 8 remains → Ten-thousands digit is 8

So when we subtract, the final answer is:

✅ **Final Answer: 85,569**

74. Add 63066 + 2387

Step 1: Write the numbers one under the other, making sure the digits are lined up by place value:

```
  63066
+ 2387
---------
```

Step 2: Start adding from the right (ones place) and move left:

- 6 + 7 = 13 → Write down 3 and carry over 1
- 6 + 8 = 14, plus the 1 carried = 15 → Write down 5 and carry over 1
- 0 + 3 = 3, plus the 1 carried = 4 → Hundreds digit is 4
- 3 + 2 = 5 → Thousands digit is 5
- 6 remains → Ten-thousands digit is 6

So when we add it all up, the final answer is:

✅ **Final Answer: 65,453**

75. **Subtract 96161 - 3342**

Step 1: Write the numbers one under the other, making sure the digits are lined up by place value:

```
  96161
-  3342
 ---------
```

Step 2: Start subtracting from the right (ones place) and move left:

- 1 - 2: We can't subtract 2 from 1, so borrow from the 6 → 11 - 2 = 9
- 5 (after borrowing) - 4 = 1 → Tens digit is 1
- 1 - 3: We can't subtract 3 from 1, so borrow from the 6 → 11 - 3 = 8
- 5 (after borrowing) - 3 = 2 → Thousands digit is 2
- 9 remains → Ten-thousands digit is 9

So when we subtract, the final answer is:

✅ **Final Answer: 92,819**

76. **Add 66709 + 2833**

Step 1: Write the numbers one under the other, making sure the digits are lined up by place value:

```
  66709
+  2833
 ---------
```

Step 2: Start adding from the right (ones place) and move left:

- 9 + 3 = 12 → Write down 2 and carry over 1
- 0 + 3 = 3, plus the 1 carried = 4 → Tens digit is 4
- 7 + 8 = 15 → Write down 5 and carry over 1
- 6 + 2 = 8, plus the 1 carried = 9 → Thousands digit is 9
- 6 remains → Ten-thousands digit is 6

So when we add it all up, the final answer is:

✅ **Final Answer: 69,542**

77. Subtract 78231 - 1828

Step 1: Write the numbers one under the other, making sure the digits are lined up by place value:

```
 78231
- 1828
---------
```

Step 2: Start subtracting from the right (ones place) and move left:

• 1 - 8: We can't subtract 8 from 1, so borrow from the 3 → 11 - 8 = 3
• 2 (after borrowing) - 2 = 0 → Tens digit is 0
• 2 - 8: We can't subtract 8 from 2, so borrow from the 8 → 12 - 8 = 4
• 7 (after borrowing) - 1 = 6 → Thousands digit is 6
• 7 remains → Ten-thousands digit is 7

So when we subtract, the final answer is:

✅ **Final Answer: 76,403**

78. Add 92249 + 8731

Step 1: Write the numbers one under the other, making sure the digits are lined up by place value:

```
 92249
+ 8731
---------
```

Step 2: Start adding from the right (ones place) and move left:

• 9 + 1 = 10 → Write down 0 and carry over 1
• 4 + 3 = 7, plus the 1 carried = 8 → Tens digit is 8
• 2 + 7 = 9 → Hundreds digit is 9
• 2 + 8 = 10 → Write down 0 and carry over 1
• 9 + 1 (carried) = 10 → Ten-thousands digit is 10

So when we add it all up, the final answer is:

✅ **Final Answer: 100,980**

79. Subtract 66617 - 1659

Step 1: Write the numbers one under the other, making sure the digits are lined up by place value:

```
 66617
- 1659
--------
```

Step 2: Start subtracting from the right (ones place) and move left:

- 7 - 9: We can't subtract 9 from 7, so borrow from the 1 → 17 - 9 = 8
- 0 (after borrowing) - 5: We can't subtract 5 from 0, so borrow from the 6 → 10 - 5 = 5
- 5 (after borrowing) - 6: We can't subtract 6 from 5, so borrow from the next 6 → 15 - 6 = 9
- 5 (after borrowing) - 1 = 4
- 6 remains → Ten-thousands digit is 6

So when we subtract, the final answer is:
Final Answer: 64,958

8[illegible] + 7423

[illegible] numbers one under the other, [illegible] digits are lined up by place [illegible]

[illegible] adding from the right (ones [illegible] left:

- [illegible] Write down 1 and carry over 1
- [illegible] the 1 carried = 9 → Tens
- [illegible]
- [illegible] Hundreds digit is 6
- [illegible] Thousands digit is 9
- [illegible] Ten-thousands digit is 1

[illegible] it all up, the final answer is:

[illegible]: 19,69

Multiplication Using Place Value

81. Multiply: 772 × 3

Step 1: Break apart the number by place value:

- Hundreds: 700
- Tens: 70
- Ones: 2

Step 2: Multiply each part by 3:

- 700 × 3 = 2100
- 70 × 3 = 210
- 2 × 3 = 6

Step 3: Add all the parts together:

- 2100 + 210 + 6 = 2316

Final Answer: 2316

82. Multiply: 362 × 3

Step 1: Break apart the number by place value:

- Hundreds: 300
- Tens: 60
- Ones: 2

Step 2: Multiply each part by 3:

- 300 × 3 = 900
- 60 × 3 = 180
- 2 × 3 = 6

Step 3: Add all the parts together:

- 900 + 180 + 6 = 1086

Final Answer: 1086

83. Multiply: 651 × 6

Step 1: Break apart the number by place value:

- Hundreds: 600
- Tens: 50
- Ones: 1

Step 2: Multiply each part by 6:

- 600 × 6 = 3600
- 50 × 6 = 300
- 1 × 6 = 6

Step 3: Add all the parts together:

- 3600 + 300 + 6 = 3906

Final Answer: 3906

84. Multiply: 935 × 2

Step 1: Break apart the number by place value:

- Hundreds: 900
- Tens: 30
- Ones: 5

Step 2: Multiply each part by 2:

- 900 × 2 = 1800
- 30 × 2 = 60
- 5 × 2 = 10

Step 3: Add all the parts together:

- 1800 + 60 + 10 = 1870

Final Answer: 1870

85. Multiply: 520 × 6

Step 1: Break apart the number by place value:

- Hundreds: 500
- Tens: 20
- Ones: 0

Step 2: Multiply each part by 6:

- 500 × 6 = 3000
- 20 × 6 = 120
- 0 × 6 = 0

Step 3: Add all the parts together:

- 3000 + 120 + 0 = 3120

Final Answer: 3120

86. Multiply: 689 × 2

Step 1: Break apart the number by place value:

- Hundreds: 600
- Tens: 80
- Ones: 9

Step 2: Multiply each part by 2:

- 600 × 2 = 1200
- 80 × 2 = 160
- 9 × 2 = 18

Step 3: Add all the parts together:

- 1200 + 160 + 18 = 1378

Final Answer: 1378

87. Multiply: 376 × 7

Step 1: Break apart the number by place value:

- Hundreds: 300
- Tens: 70
- Ones: 6

Step 2: Multiply each part by 7:

- 300 × 7 = 2100
- 70 × 7 = 490
- 6 × 7 = 42

Step 3: Add all the parts together:

- 2100 + 490 + 42 = 2632

Final Answer: 2632

88. Multiply: 805 × 7

Step 1: Break apart the number by place value:

- Hundreds: 800
- Tens: 0
- Ones: 5

Step 2: Multiply each part by 7:

- 800 × 7 = 5600
- 0 × 7 = 0
- 5 × 7 = 35

Step 3: Add all the parts together:

- 5600 + 0 + 35 = 5635

Final Answer: 5635

89. Multiply: 568 × 5

Step 1: Break apart the number by place value:

- Hundreds: 500
- Tens: 60
- Ones: 8

Step 2: Multiply each part by 5:

- 500 × 5 = 2500
- 60 × 5 = 300
- 8 × 5 = 40

Step 3: Add all the parts together:

- 2500 + 300 + 40 = 2840

Final Answer: 2840

90. Multiply: 588 × 5

Step 1: Break apart the number by place value:

- Hundreds: 500
- Tens: 80
- Ones: 8

Step 2: Multiply each part by 5:

- 500 × 5 = 2500
- 80 × 5 = 400
- 8 × 5 = 40

Step 3: Add all the parts together:

- 2500 + 400 + 40 = 2940

Final Answer: 2940

91. Multiply: 817 × 4

Step 1: Break apart the number by place value:

- Hundreds: 800
- Tens: 10
- Ones: 7

Step 2: Multiply each part by 4:

- 800 × 4 = 3200
- 10 × 4 = 40
- 7 × 4 = 28

Step 3: Add all the parts together:

- 3200 + 40 + 28 = 3268

Final Answer: 3268

92. Multiply: 197 × 2

Step 1: Break apart the number by place value:

- Hundreds: 100
- Tens: 90
- Ones: 7

Step 2: Multiply each part by 2:

- 100 × 2 = 200
- 90 × 2 = 180
- 7 × 2 = 14

Step 3: Add all the parts together:

- 200 + 180 + 14 = 394

Final Answer: 394

… × 5

…art the number by place

…100

…each part by 5:

…500

…0

…15

…he parts together:

…+ 15 = 565

…565

…69 × 3

…art the number by place

…800

…each part by 3:

… = 2400

…180

…7

…he parts together:

…80 + 27 = 2607

…2607

95. Multiply: 785 × 7

Step 1: Break apart the number by place value:

- Hundreds: 700
- Tens: 80
- Ones: 5

Step 2: Multiply each part by 7:

- 700 × 7 = 4900
- 80 × 7 = 560
- 5 × 7 = 35

Step 3: Add all the parts together:

- 4900 + 560 + 35 = 5495

Final Answer: 5495

96. Multiply: 588 × 4

Step 1: Break apart the number by place value:

- Hundreds: 500
- Tens: 80
- Ones: 8

Step 2: Multiply each part by 4:

- 500 × 4 = 2000
- 80 × 4 = 320
- 8 × 4 = 32

Step 3: Add all the parts together:

- 2000 + 320 + 32 = 2352

Final Answer: 2352

97. Multiply: 184 × 8

Step 1: Break apart the number by place value:

- Hundreds: 100
- Tens: 80
- Ones: 4

Step 2: Multiply each part by 8:

- 100 × 8 = 800
- 80 × 8 = 640
- 4 × 8 = 32

Step 3: Add all the parts together:

- 800 + 640 + 32 = 1472

Final Answer: 1472

98. Multiply: 201 × 8

Step 1: Break apart the number by place value:

- Hundreds: 200
- Tens: 0
- Ones: 1

Step 2: Multiply each part by 8:

- 200 × 8 = 1600
- 0 × 8 = 0
- 1 × 8 = 8

Step 3: Add all the parts together:

- 1600 + 0 + 8 = 1608

Final Answer: 1608

99. Multiply: 650 × 7

Step 1: Break apart the number by place value:

- Hundreds: 600
- Tens: 50
- Ones: 0

Step 2: Multiply each part by 7:

- 600 × 7 = 4200
- 50 × 7 = 350
- 0 × 7 = 0

Step 3: Add all the parts together:

- 4200 + 350 + 0 = 4550

Final Answer: 4550

100. Multiply: 466 × 8

Step 1: Break apart the number by place value:

- Hundreds: 400
- Tens: 60
- Ones: 6

Step 2: Multiply each part by 8:

- 400 × 8 = 3200
- 60 × 8 = 480
- 6 × 8 = 48

Step 3: Add all the parts together:

- 3200 + 480 + 48 = 3728

Final Answer: 3728

25 Multi-Step 4th Grade Story Problems with Step-by-Step Solutions

1. Kalel collected 407,863 stamps and Whitley collected 481,720 stamps. Who collected more? By how much?

Step 1: Compare numbers. Whitley collected more.
Step 2: Subtract: 481,720 - 407,863 = 73,857
Answer: Whitley collected 73,857 more stamps.

2. The population of the town Boggle-bee-bear is 38,723. What is the population rounded to the nearest thousand and nearest ten thousand?

To nearest 1,000: 39,000 (since $723 \geq 500$)
To nearest 10,000: 40,000 (since $8{,}723 \geq 5{,}000$)
Answer: 39,000 and 40,000.

3. A factory produced 29,176 toys in January and 27,447 toys in February. If 6,082 toys were defective and thrown out, how many good toys were produced? If each toy cost 1.25 dollars to make, how much money did the factory lose?

Step 1: Add total production: 29,176 + 27,447 = 56,623
Step 2: Subtract defective: 56,623 - 6,082 = 50,541
Step 3: Multiply loss: 6,082 × 1.25 = $7,602.50
Answer: 50,541 good toys; $7,602.50 lost.

4. Each school supply box contains 571 pencils. If a [illegible] boxes, how many pencils will they receive in total? 10% of the pencils they rec[illegible] are defective. How many good pencils are there?

Step 1: Total pencils: $571 \times 6 = 3{,}426$

Step 2: Find 10%: $342.6 \rightarrow 343$

Step 3: Subtract: $3{,}426 - 343 = 3{,}083$

Answer: 3,083 good pencils.

5. Evelynn has 578 new books to arrange evenly on [illegible] many books will be on each shelf? Will any be left over?

Step 1: Divide: $578 \div 5 = 115$ R3

Answer: 115 books per shelf; 3 left over.

6. Sophia collected 225,728 stamps and Gabriella [illegible]53 stamps. Who collected more? By how much?

Step 1: Compare—Gabriella has more.

Step 2: Subtract: $226{,}353 - 225{,}728 = 625$

Answer: Gabriella collected 625 more.

7. The population of the town Snicker-dee-doodle is [illegible] is the population rounded to the nearest thousand and nearest ten thousand?

To nearest 1,000: 58,000

To nearest 10,000: 60,000

Answer: 58,000 and 60,000.

8. A factory produced 48,744 pop-it toys in January and 12,274 toys in February. If 9,056 pop-it toys were defective and thrown out, how many good pop-it toys were produced?

Step 1: Add: 48,744 + 12,274 = 61,018

Step 2: Subtract defective: 61,018 - 9,056 = 51,962

Answer: 51,962 good toys.

9. Each school supply box contains 603 markers. If a school orders 2 boxes, how many markers will they receive in total?

Step 1: Multiply: 603 × 2 = 1,206

Answer: 1,206 markers.

10. A bank teller has 583 stacks of money to arrange evenly on 7 shelves. How many stacks of money will be on each shelf? Will any be left over?

Step 1: Divide: 583 ÷ 7 = 83 R2

Answer: 83 per shelf; 2 left over.

11. Harbor collected 459,944 energy drinks and Aubrey collected 314,682 energy drinks. Who collected more? By how much?

Step 1: Compare—Harbor has more

Step 2: Subtract: 459,944 - 314,682 = 145,262

Answer: Harbor collected 145,262 more.

12. The population of dogs in the town of Ruff-Woof-Ruff is 99,618. What is the population of dogs rounded to the nearest thousand and nearest ten thousand?

To nearest 1,000: 100,000

To nearest 10,000: 100,000

Answer: 100,000 and 100,000.

13. A factory produced 452 jelly-cats in April that sold at $150 each and 351 jelly-cats in May that were limited edition and sold for $200 each. How much money did the factory make in April and May? Which month made more?

April: 452 × $150 = $67,800

May: 351 × $200 = $70,200

Answer: $67,800 in April, $70,200 in May; May made more.

14. Each lunch box contains 3 mini crumble cookies. If a school orders 855 boxes, how many mini crumble cookies will they receive in total?

Step 1: Multiply: 3 × 855 = 2,565

Answer: 2,565 cookies.

15. A grocery worker has 861 boxes of cereal to arrange evenly on 5 shelves. How many boxes will be on each shelf? Will any be left over?

Step 1: Divide: 861 ÷ 5 = 172 R1

Answer: 172 per shelf; 1 left over.

16. Bob collected 737,602 toy cars and Victor collected 582,744 toy cars. Who collected more? By how much?

Step 1: Compare—Bob has more

Step 2: Subtract: 737,602 - 582,744 = 154,858

Answer: Bob collected 154,858 more.

17. The population of cats in a town is 24,030. What is the population rounded to the nearest thousand and nearest ten thousand?

To nearest 1,000: 24,000

To nearest 10,000: 20,000

Answer: 24,000 and 20,000.

18. Cane's Fast-Food Chicken produced 42,262 chicken tenders in January and 31,082 in February. If 11,044 tenders were defective, how many were good?

Step 1: Add: 42,262 + 31,082 = 73,344

Step 2: Subtract: 73,344 - 11,044 = 62,300

Answer: 62,300 good tenders.

19. Each bag of strawberry seeds contains 451 seeds. If a school orders 6 boxes, how many total seeds are there? If the garden has 12 pots that take 125 seeds each, how many are left over?

Total seeds: 451 × 6 = 2,706

Planted: 12 × 125 = 1,500

Left: 2,706 - 1,500 = 1,206

Answer: 1,206 seeds left.

20. Your mom asked you to organize 602 cups on 9 [illegible] any cups per shelf? Any left over?

Step 1: Divide: 602 ÷ 9 = 66 R8

Answer: 66 per shelf; 8 left over.

21. Mom found 876 toys on the floor. She told her chi[illegible] all the toys, and for each toy she found she was going to charge them $1.50 [illegible] up some and left 789. How many toys did they pick up? How much mone[illegible] Mom for picking up their toys?

Step 1: Subtract: 876 total toys - 789 left over toys = [illegible]

Step 2: Times: 789 x $1.50= $1, 183.50

Answer: 87 toys picked up; $1, 183.50 is owed to M[illegible]

22. Gabriella watched 41,282 TikTok videos. What is [illegible] the nearest thousand and ten thousand?

To nearest 1,000: 41,000

To nearest 10,000: 40,000

Answer: 41,000 and 40,000.

23. A paint factory made 24,089 cans in August an[illegible]ember. If 9,636 were defective and each cost $2.50 to make, how muc[illegible]st?

Total: 24,089 + 13,693 = 37,782

Good: 37,782 - 9,636 = 28,146

Loss: 9,636 × $2.50 = $24,090

Answer: 28,146 good; $24,090 lost.

24. Vic gives each of his 5 children $250 for school. If each spends 75%, how much is left per child?

Spent: 250 × 0.75 = 187.50

Left = 250 - 187.50 = 62.50

Answer: $62.50 left per child.

25. Robbie has 76 medals and 3 baskets. How many per basket? Will there be any left over?

Step 1: Divide: 76 ÷ 3 = 25 R1

Answer: 25 per basket; 1 left over.

Answer Key - 4th Grade Math Post-Test (Challenging)

1. Write the number nine hundred two thousand, seventy-five in standard form:

Answer: 902075

2. Round 97426 to the nearest ten thousand:

97426 → Look at the thousands digit (7)

Since it's 5 or more, round up: 100000

3. Compare using >, <, or =: 100491 ___ 99410

Answer: 100491 > 99410

4. What is the value of the digit 7 in the number 874309?

Answer: 7000 (7 is in the thousands place)

5. Add: 72938 + 18426

Step-by-step:

```
  72938
+ 18426
  ---------
  91364
```

Answer: 91364

6. Subtract: 91023 - 4729

Step-by-step:

```
  91023
-  4729
 ---------
  86294
```

Answer: 86294

7. Multiply: 73 × 48

Step-by-step:

$73 \times 48 = (70 + 3) \times 48 = (70\times48) + (3\times48) = 3360 + 144 = 3504$

Answer: 3504

8. Estimate the difference between 85390 and 29456:

Round to nearest thousand: 85400 - 29400 = 56000

Answer: About 56000

9 Word Problem:

625 toys/day × 7 days = 4375 toys

4375 - 3590 = 785 toys left

Answer: 785

10. Divide: 935 ÷ 5 – (BONUS QUESTION!)

Step 1: Set up the long division.

We're dividing 935 by 5.

__187__

5|935

Step 2: Divide the hundreds place (9).

5 goes into 9 one time. Write 1 above the 9.

Subtract: 9 - 5 = 4

Bring down the 3 to make 43.

Step 3: Divide the tens place (43).

5 goes into 43 eight times. Write 8 above the 3.

Subtract: 43 - 40 = 3

Bring down the 5 to make 35.

Step 4: Divide the ones place (35).

5 goes into 35 seven times. Write 7 above the 5.

Subtract: 35 - 35 = 0. No remainder!

Final Answer: 187

Certificate of Completion

This is proudly awarded to:

____

For successfully completing the

Math Practice Workbook: 4th Grade

Numbers & Operations in Base Ten

This student has shown strong understanding of place value, multi-digit addition and subtraction, and strategies for multiplication and division.

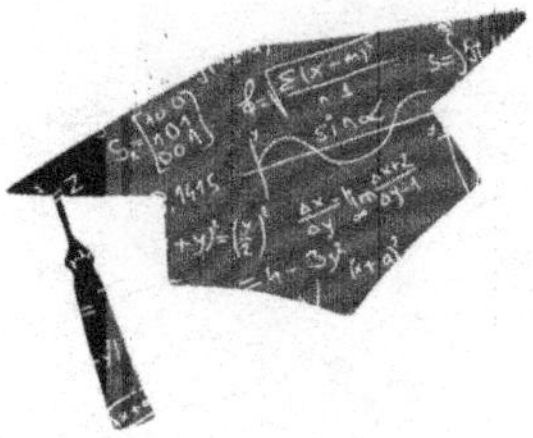

Completed on: ________________________________

Signed: ____________________________________

(Parent/Teacher)

I GOT THIS!
I GOT THIS!
I GOT THIS!

Keep up the great work and continue your math journey! 🧠✨

Say it with me, "I Got This!"

Say it with me, "I Got This!"

Say it with me, "I Got This!"

Say it with me, "I Got This!"

Say it with me, "I Got This!"

Say it with me, "I Got This!"

Say it with me, "I Got This!"

Say it with me, "I Got This!"

Say it with me, "I Got This!"

Say it with me, "I Got This!"

Say it with me, “I Got This!”

Say it with me, "I Got This!"

Say it with me, "I Got This!"

Say it with me, "I got this!"

Say it with me, "I Got This!"

Say it with me, "I Got This!"

Say it with me, "I Got This!"

Say it with me, "I Got This!"

Copyright & Disclaimer

Educational Purpose Only

This workbook was carefully crafted to help 4th grade students build confidence and develop strong math skills in a fun and approachable way. While every effort has been made to ensure accuracy, Chelse's Creations makes no guarantees regarding outcomes and is not responsible for errors or omissions.

Thank You

Thank you for supporting creative educators and resources made with care. By using this workbook, you are encouraging learning, imagination, and growth. A portion of all proceeds goes back into the community through our volunteer organization, Helping from the Heart Foundation. Thank you for making a difference!

For more learning resources, visit: **Chelse's Creations Amazon Store**

Chelse Burton is an accomplished mom, nurse, author, and artist who is passionate about education and helping others. Her diverse talents and dedication shine through everything she creates—from inspiring novels and educational children's books to beautiful artwork and even her own line of homemade seasonings!

With a background as a trauma nurse, educator, and case manager, Chelse has seen both the strength and struggles of people in her community. These experiences motivate her to share messages of hope, learning, and creativity through her work.

Chelse's passion for education shines brightest in her children's books and educational series, "I Got This!". Through engaging stories and fun workbooks, she encourages young learners to explore, think deeply, and believe in themselves. Her mission is to inspire curiosity and confidence in every child—helping them see that learning can be both exciting and empowering!

Chelse and her husband Victor, a proud Marine veteran, are the founders of **Helping from the Heart Foundation**, a volunteer organization dedicated to serving the community with the same dedication, integrity, and leadership that defines the Marine Corps. The foundation is committed to making a lasting impact through volunteer efforts that support those in need—especially fellow veterans, first responders, and underserved communities.

A portion of all proceeds made from Chelse's Creations, including this workbook, goes back into the community through the volunteer organization, Helping from the Heart Foundation. ***Thank you* for making a difference!**

Made in the USA
Monee, IL
05 November 2026

67386640R00057